G. L. Abu Youcef

Preventing RisksLié aux Pesticides

G. L. Abu Youcef

Preventing RisksLié aux Pesticides

ScienciaScripts

Imprint
Any brand names and product names mentioned in this book are subject to trademark, brand or patent protection and are trademarks or registered trademarks of their respective holders. The use of brand names, product names, common names, trade names, product descriptions etc. even without a particular marking in this work is in no way to be construed to mean that such names may be regarded as unrestricted in respect of trademark and brand protection legislation and could thus be used by anyone.

Cover image: www.ingimage.com

This book is a translation from the original published under ISBN 978-620-6-72447-6.

Publisher:
Sciencia Scripts
is a trademark of
Dodo Books Indian Ocean Ltd. and OmniScriptum S.R.L publishing group

120 High Road, East Finchley, London, N2 9ED, United Kingdom
Str. Armeneasca 28/1, office 1, Chisinau MD-2012, Republic of Moldova, Europe
Managing Directors: Ieva Konstantinova, Victoria Ursu
info@omniscriptum.com

Printed at: see last page
ISBN: 978-620-8-61515-4

Preventing risks linked to pesticides

Image: iStock

By: G.L/ABU YOUCEF

Table of contents

Book summary

"Prévenir les Risques Liés aux Pesticides" is an exhaustive exploration of the dangers associated with the use of pesticides, both for human health and the environment. At a time when agricultural practices are having to evolve in response to ecological challenges, this book aims to raise readers' awareness of the often little-known impacts of plant protection products. Each chapter deals with a specific aspect, from the types of pesticide to viable alternatives, including regulations and the training required for safe use.

The book begins with an introduction to pesticides, providing a historical and scientific overview of their use. It then details the various health risks, including the long-term effects on the health of farmers and consumers. At the same time, the environmental consequences, such as soil and water contamination, are also highlighted, underlining the importance of an integrated approach to pest management.

The book proposes practical solutions to reduce these risks, focusing on sustainable farming practices and alternatives to chemical pesticides. Case studies illustrate successful initiatives, and practical advice is provided to help farmers adopt less harmful methods.

Finally, the book concludes by calling for collaboration between farmers, consumers and decision-makers. By adopting a collective approach, it is possible to face up to the challenges posed by the use of pesticides and guarantee a healthy future for our planet and its inhabitants. This book is not just a guide; it is a manifesto for urgent and necessary change.

Author:
G.L/ABU
Supervisor, trainer and doctoral student in SST

A word from the author

Dear readers,

In our modern era, agriculture has often been synonymous with increased productivity, technological innovation and the ability to feed an ever-growing world population. However, this quest for higher yields must not obscure the crucial issues raised by the use of pesticides. These chemical substances, although effective in combating crop pests and diseases, present major risks to human health and the environment. It is essential to be aware of these dangers so that we can make informed choices that protect not only our well-being, but also that of our planet.

This book is not simply a collection of information, but a real call to awareness and collective action. As the author, my mission is to provide you with an in-depth understanding of the risks associated with pesticides. Each chapter deals with a specific aspect of this complex issue, from the harmful effects on health to the degradation of our ecosystems. My hope is that you will be able to grasp the scale of these issues and the implications they have for our daily lives.

At the same time, it is just as important to put forward practical solutions to mitigate these risks. This book proposes viable alternatives to chemical pesticides, as well as sustainable agricultural practices that respect our health and our environment. I wanted each section to be enriching and practical, offering advice and strategies that every reader can apply, whether you are a farmer, a committed consumer or simply curious to better understand this subject.

It is crucial to recognise that change cannot be achieved without active collaboration between all the players in the food chain. This book is aimed at everyone: farmers, political decision-makers, researchers and consumers. Together, we have the capacity to transform our farming practices and promote methods that protect our health and our environment. It's a collective effort that requires the participation of everyone, because the challenges we face today demand shared, innovative solutions.
Finally, I encourage you to approach this book with an open mind and a

willingness to learn. Each chapter has been designed to help you navigate this complex subject and encourage discussion about the future of our agriculture. By taking a proactive approach, we can make a difference and ensure a healthy future for generations to come. Thank you for reading, and let's work together for meaningful and lasting change.

With kindness,

Author:
G.L/ABU
Supervisor, trainer and doctoral student in SST

-Introduction to Pesticides

Pesticides are substances used to prevent, repel or eliminate harmful organisms that compromise the health of crops. They play a crucial role in modern agriculture, helping to protect crops and increase productivity. However, their use raises important questions about human health and the environment, making it necessary to explore their nature and impact in depth.

Image: iStock

Pesticides fall into several categories, including insecticides, herbicides, fungicides and rodenticides. Each category targets a specific type of pest. For example, insecticides are designed to control insect pests, while herbicides are used to control weeds. This specialisation allows farmers to select products tailored to their needs, but it also entails risks of undesirable side-effects.

One of the major concerns about pesticides is their impact on human health. Numerous studies have linked exposure to pesticides to a range of health problems, from neurological disorders to cancer. Farmers, in particular, are often the most exposed, but pesticide residues can also be found in food and water, affecting the whole population. It is therefore essential to understand not only the benefits of these products, but also the precautions to be taken to minimise the risks.

The environmental impact of pesticides is just as worrying. Their use can lead to soil and water contamination, affecting not only flora and fauna, but ecosystems as a whole. Pesticides can disrupt food chains and lead to the decline of certain species, thus contributing to the loss of biodiversity. This fact underlines the importance of sensible pesticide management in protecting our natural resources.

In response to these challenges, regulations have been introduced to govern the use of pesticides. Various countries have passed laws aimed at limiting the risks associated with their application, by imposing safety standards and encouraging the use of alternative methods. These developments are crucial to ensuring that farming practices respect both human health and the environment.

Finally, educating and raising awareness among farmers, consumers and the general public is essential to promoting the responsible use of pesticides. By understanding the associated risks and exploring sustainable alternatives, everyone can contribute to a healthier and more environmentally-friendly agricultural future. This introduction sets the scene for a more in-depth discussion of the issues surrounding pesticides, highlighting the need to strike a balance between agricultural productivity and the protection of our health and our planet.

-History of Pesticide Use

The use of substances to protect crops dates back to ancient times. Ancient civilisations, such as the Egyptians, used natural mixtures of plants and minerals to repel harmful insects. For example, plant resins and extracts were applied to crops to protect them from pests. These traditional methods, although rudimentary, bear witness to an early awareness of the importance of crop protection.

Over the centuries, with the advent of modern chemistry in the 19th century, industrialisation led to the synthesis of new chemicals. In 1867, the first synthetic insecticide, iron sulphate, was introduced onto the market. However, it was in the 1940s, with the discovery and marketing of the insecticide DDT (dichloro-diphenyl-trichloroethane), that the use of pesticides really exploded. DDT was widely used to combat disease-carrying insects, such as malaria, and was hailed for its effectiveness (Carson, 1962).

The 1960s and 1970s marked a period of growing awareness of the risks associated with the use of pesticides, particularly after the publication of "Silent Spring" by Rachel Carson. This book highlighted the dangers of DDT and other pesticides on wildlife and human health, sparking a worldwide reaction and initiating environmental movements (Carson, 1962). Governments began to introduce regulations to control the use of pesticides, leading to bans on certain toxic products.

In the decades that followed, alternatives to chemical pesticides were sought. Integrated pest management (IPM) methods, which combine several pest management approaches, were developed to reduce dependence on pesticides. This approach aims to use natural predators, cultural techniques and other methods to control pest populations in a more sustainable way (Gurr et al., 2016). However, pesticides continue to play a major role in modern agriculture, particularly in developing countries where resources to adopt alternative methods are limited.

The rise of biotechnology in the 1990s also transformed the pesticide landscape. The development of pest-resistant genetically modified (GM) crops has made it possible to reduce the use of certain pesticides while maintaining high yields. However, this development has sparked heated debates about the safety of GMOs and their environmental impacts (Pardo et al., 2018). Concerns about pest resistance, loss of biodiversity and effects on public health remain.

Today, the use of pesticides is subject to strict regulations in many countries, but the challenges remain. Issues of sustainability, food safety and environmental protection continue to be at the heart of discussions on agriculture. Scientists and decision-makers are seeking to strike a balance between the need to protect crops and the need to safeguard human health and the environment.

In conclusion, the history of pesticide use is marked by undeniable successes in terms of crop protection, but also by growing challenges linked to their impact on health and the environment. These developments underline the importance of a considered and sustainable approach to pesticide management, in order to ensure an agricultural future that is both productive and respectful of our planet.

References
Carson, R. (1962). Silent Spring. Houghton Mifflin.
Gurr, G. M., Wratten, S. D., & Luna, J. M. (2016). Biological Control: A Global Perspective. CABI.
Pardo, A., et al. (2018). "Impact of Genetically Modified Organisms on Biodiversity." Nature Sustainability, 1(1), 45-54.

-Types of Pesticides and How They Work

Pesticides are a diverse category of chemical substances designed to control harmful organisms. They are classified into several types, each with a specific target and a particular mode of action . This diversity allows farmers to choose the most appropriate products to protect their crops while minimising the impact on the environment.

1. Insecticides

Insecticides are pesticides designed to eliminate or control harmful insects. They can act by contact, ingestion or inhalation, disrupting the insects' nervous systems. The most commonly used insecticides include organophosphates and pyrethroids. Organophosphates inhibit enzymes essential to insect metabolism, while pyrethroids mimic a natural toxin found in certain flowers (Matsumura, 1985). However, their excessive use has led to resistance problems in certain insect species, making it necessary to look for alternative solutions.

Image: iStock

2. Herbicides

Herbicides are used to control weeds, which can compete with crops for nutrients, water and light. They generally work by disrupting essential biological processes in plants. Herbicides can be selective, targeting only certain plant species, or non-selective, killing all vegetation to which they are applied. Glyphosates, for example, inhibit a specific metabolic pathway present only in plants and certain micro-organisms, making them effective at eliminating weeds while being less toxic to animals (Dill et al., 2010).

3. Fungicides

Fungicides are substances designed to combat pathogenic fungi that can cause plant diseases. They work by inhibiting the growth of fungi or by killing them directly. Fungicides can be systemic, moving inside the plant, or contact, acting only on the surface. Triazoles and strobilurins are two commonly used classes of fungicide. Triazoles inhibit the biosynthesis of sterols, which are essential for fungal cell membranes, while strobilurins interfere with cell respiration (Fravel, 2005).

4. Rodenticides

Rodenticides are designed to control rodents, which can cause significant damage to crops and infrastructure. These substances generally act by disrupting essential biological functions, often by inhibiting blood coagulation or affecting the nervous system. Anticoagulants such as bromadiolone are among the most widely used rodenticides, causing internal bleeding in rodents after ingestion (Mason et al., 2004).

5. Biological pesticides

Biological pesticides, or biopesticides, include biological control agents such as viruses, bacteria and fungi that naturally control pest populations. They offer a less toxic alternative to chemical pesticides and are often more targeted. For example, Bacillus thuringiensis (Bt) is a bacterium that produces toxins specific to insects, enabling natural control without harming other organisms (Liu et al., 2005).

6. Systemic and Contact Pesticides

Pesticides can also be classified according to their mode of action. Systemic pesticides are absorbed by plants and transported into their tissues, offering prolonged protection against pests. In contrast, contact pesticides act only on the surfaces to which they are applied, requiring more frequent application to maintain efficacy. Understanding these differences is essential if pesticides are to be applied effectively and responsibly.

In summary, there are many different types of pesticide, each with specific mechanisms of action that target different pests. Although these substances play a crucial role in crop protection, their use must be carefully managed to minimise the impact on human health and the environment. Continuous research into alternatives and integrated management methods is essential to ensure sustainable agriculture.

References

Dill, G. M., et al. (2010). "Glyphosate-Resistant Crops: History, Status and Future." Pest Management Science, 66(3), 307-318.

Fravel, D. R. (2005). "Commercialization of Biocontrol." Annual Review of Phytopathology, 43, 341-356.

Liu, Y. B., et al (2005). "Bacillus thuringiensis: A Source of Insecticidal Proteins." Nature Biotechnology, 23(3), 289-293.

Matsumura, F. (1985). Toxicology of Insecticides. Plenum Press.

Mason, G. J., et al (2004). "Rodenticides: A Review of Their Use and Impact." Wildlife Society Bulletin, 32(1), 92-96.

-Risks to human health

The use of pesticides in modern agriculture has led to growing concerns about their impact on human health. Although these substances are designed to target harmful organisms, their application can have undesirable effects on the health of farm workers, consumers and people living near treated areas. Understanding these risks is essential to promote responsible use and inform food safety policies.

1. Occupational exposure

Agricultural workers are often the most exposed to pesticides, due to direct contact during application, mixing or storage. Exposure can occur through inhalation, skin contact or accidental ingestion. Studies have shown that farmers exposed to insecticides, herbicides or fungicides have an increased risk of developing health problems, including neurological disorders, respiratory diseases and cancer (Alavanja et al., 2003). Awareness-raising and training in good safety practices are essential to minimise these risks.

2. Long-term health effects

Chronic exposure to pesticides is of particular concern, as many harmful effects may not be immediately apparent. Research has established links between long-term exposure to certain pesticides and serious illnesses such as Parkinson's disease, specific types of cancer (such as non-Hodgkin's lymphoma) and endocrine disorders (Kamel et al., 2007). These illnesses can result from an accumulation of pesticides in the body, underlining the importance of ongoing monitoring of the health of agricultural workers.

3. Residues in food

Pesticide residues can also be found in the food we eat, representing a health risk for consumers. Residue levels are often regulated, but there are concerns about the safety of these limits, especially when pesticides are used intensively. Studies have shown that regular consumption of fruit and vegetables treated with pesticides can have cumulative effects on health, particularly in children and pregnant women, who are more sensitive to endocrine disruption (Bogdan et al., 2013).

4. Effects on Sensitive Populations

Certain populations, such as children, pregnant women and the elderly, are particularly vulnerable to the harmful effects of pesticides. Children, because of their still-developing metabolism and smaller size, are more likely to suffer harmful effects even with exposure to low levels of pesticides. Studies have shown that prenatal exposure to pesticides is linked to neurological and behavioural effects in children (Raanan et al., 2015). This highlights the need for strict regulation and risk assessment to protect these vulnerable groups.

5. Psychological effects

Exposure to pesticides is not limited to physiological effects; it can also have psychological consequences. Studies have shown that agricultural workers exposed to pesticides often develop symptoms of depression and anxiety, which may result from a combination of occupational stress and the neurotoxic effects of pesticides (Matsumura, 1985). The mental health of agricultural workers is an often neglected but crucial aspect that deserves particular attention.

6. Resistance and Implications for Health

The repeated and intensive use of pesticides can also lead to pest resistance, resulting in an increase in the use of more powerful and potentially more toxic chemicals for human health. This spiral of pesticide dependency poses challenges not only for the health of workers, but also for global food security. Combating resistance requires an integrated approach, combining different pest control methods to reduce reliance on chemical pesticides.

Finally, the risks to human health associated with the use of pesticides are varied and complex. From occupational exposure to food residues, via the effects on sensitive populations, these issues require increased attention. Promoting sustainable agricultural practices, strengthening regulations and raising awareness among farmers and consumers are crucial steps in protecting public health and ensuring a healthy agricultural future.

References
Alavanja, M. C. R., et al. (2003). "Pesticides and Cancer." Cancer Causes & Control, 14(2), 147-157.
Bogdan, A., et al. (2013). "Pesticide Residues in Food: A Review of the Health Risks." Journal of Food Safety, 33(2), 203-215.
Kamel, F., et al. (2007). "Neurologic and Cognitive Function in Agricultural Workers." Environmental Health Perspectives, 115(1), 103-108.
Matsumura, F. (1985). Toxicology of Insecticides. Plenum Press.
Raanan, R., et al. (2015). "Prenatal Exposure to Pesticides and Child Neurodevelopment." Environmental Health Perspectives, 123(9), 991-997.

-Environmental risks

The use of pesticides in modern agriculture poses significant risks to the environment, affecting not only biodiversity but also soil and water quality. Although these substances are designed to protect crops, their impact on ecosystems is often complex and can have harmful long-term consequences. Understanding these risks is essential if we are to promote sustainable agricultural practices.

1. Soil contamination

Pesticides can accumulate in the soil, affecting its health and fertility. When these chemicals are applied, some of them can be absorbed by plants, but others can seep into the soil and persist. This can disrupt the microbial communities essential for the decomposition of organic matter and soil fertility. Studies show that some pesticides can reduce microbial diversity, with implications for soil quality and its ability to support sustainable agriculture (Giller et al., 2009).

2. Water pollution

The contamination of surface water and groundwater is another major concern linked to the use of pesticides. When it rains, pesticides applied to crops can be washed into streams, rivers and lakes. This aquatic pollution can have devastating effects on aquatic ecosystems, leading to the death of fish and other aquatic organisms . Studies have detected pesticide residues in sources of drinking water, posing a risk to human health (Gilliom et al., 2006).

3. Impact on biodiversity

The use of pesticides can lead to a reduction in biodiversity, both in terms of plant and animal species. Non-selective pesticides, which kill a wide range of organisms, can affect pollinators such as bees, as well as other beneficial insects. The loss of pollinators has serious consequences for food production and ecosystem health (Potts et al., 2010). In addition, declining insect populations can disrupt food chains, leading to cascading effects in ecosystems.

4. Pest resistance

Repeated use of pesticides can also lead to pest resistance, making treatments less effective and necessitating the application of more powerful or more toxic chemicals. This spiral of pesticide dependency not only increases environmental risks, but can also compromise food security. Pest resistance is a growing challenge that requires an integrated approach to pest management, combining biological and cultural methods to reduce reliance on chemical pesticides (Gurr et al., 2016).

5. Effects on wildlife

Pesticides can have harmful effects on wildlife, particularly birds, mammals and amphibians. Exposure to pesticides can lead to reproductive problems, altered behaviour and mortality. For example, studies have shown that certain insecticides can cause deformities in amphibians and affect bird reproduction (Mineau and Whiteside, 2013). Reduced wildlife populations can also contribute to biodiversity loss and unbalanced ecosystems.

Image : Freepik

6. Long-term effects

The long-term effects of pesticide use on the environment are often difficult to assess. Pesticide residues can persist in soil and water for years, affecting the health of ecosystems long after they have been applied. In addition, the complex interactions between different pesticides and other pollutants can lead to synergistic effects, making it difficult to predict the environmental consequences (Baker et al., 2013). This uncertainty underlines the importance of rigorous environmental assessment before pesticides are used.

In short, the environmental risks associated with the use of pesticides are numerous and complex. From soil and water contamination to the impact on biodiversity and wildlife, these issues require special attention and proactive management. Promoting sustainable farming practices and encouraging research into alternatives to chemical pesticides are crucial steps in protecting our environment and ensuring a sustainable future for agriculture.

References

Baker, R., et al. (2013). "Long-term Ecological Effects of Pesticides." Environmental Science & Technology, 47(12), 6357-6369.

Gilliom, R. J., et al. (2006). "Pesticides in the Nation's Streams and Ground Water, 1992-2001." US Geological Survey Circular 1291.

Giller, K. E., et al. (2009). "Reducing Pesticide Use in Agriculture: A Review." Agriculture, Ecosystems & Environment, 133(4), 243-251.

Gurr, G. M., Wratten, S. D., & Luna, J. M. (2016). Biological Control: A Global

Perspective. CABI.
Mineau, P., & Whiteside, M. (2013). "Pesticide Risk to Birds: A Review and Assessment of the Evidence." Environmental Pollution, 173, 229-231.
Potts, S. G., et al. 2010. "Global Pollinator Declines: Trends, Impacts and Drivers." Trends in Ecology & Evolution, 25(6), 345-353.

-Pesticide regulations and standards

The use of pesticides is strictly regulated in many parts of the world, particularly in Europe and the United States. These regulations aim to protect human health, wildlife and the environment, while providing farmers with access to effective pest management tools. However, approaches and standards vary considerably between these two regions.

1. Regulations in Europe

In Europe, the regulation of pesticides is mainly governed by Regulation (EC) 1107/2009, which lays down harmonised rules for the marketing of plant protection products. The aim of this regulation is to ensure that only safe and effective pesticides are authorised. Risk assessment is a rigorous process that takes into account the impact on human health, the environment and biodiversity. Member countries must comply with these standards, but they also have the option of adopting stricter measures.

The European Food Safety Authority (EFSA) plays a key role in the assessment of pesticides. It assesses applications for marketing authorisation and provides scientific opinions on product safety. In addition, European legislation imposes maximum residue limits (MRLs) in food and drink, ensuring that pesticide levels in food products do not exceed thresholds considered safe (European Commission, 2019).

Another important aspect of European regulations is the periodic review of pesticide authorisations. This means that products that have already been authorised can be re-evaluated in the light of new scientific data, allowing those deemed obsolete or dangerous to be eliminated. In addition, the European Union has banned the use of certain pesticides, such as neonicotinoid, due to their harmful effects on pollinators (European Commission, 2018).

2. Regulations in the United States

In the United States, pesticide regulation is primarily governed by the Federal Insecticide, Fungicide and Rodenticide Act (FIFRA), administered by the Environmental Protection Agency (EPA). FIFRA requires all pesticides to be registered with the EPA before they are placed on the market. The EPA assesses the potential risks to human health and the environment before granting approval. This assessment includes studies on exposure, toxicity and long-term effects (EPA, 2021).

The EPA also sets maximum residue limits (MRLs) for food products, similar to those in the EU. However, the approval process may be less rigorous than in Europe. In addition, some states have the ability to impose stricter regulations than those of the federation, which can create disparities in the application of standards across the country.

An important aspect of US regulation is the Pesticide Re-evaluation Programme, which aims to ensure that products remain safe over time. The EPA regularly re-evaluates pesticides to ensure that they continue to meet safety standards. However, this re-evaluation can take several years, raising concerns about the speed with which potentially dangerous products can be withdrawn from the market.

3. Comparison of Approaches

The main difference between regulations in Europe and the United States lies in the rigour of pesticide assessment and re-evaluation procedures. Europe adopts a more precautionary approach, with stricter standards for pesticide authorisation and a tendency to ban substances deemed dangerous. In the United States, on the other hand, the assessment process is more focused on risk management, which may allow certain products to remain on the market for longer even though they present safety concerns.

4. Impact of regulations

Pesticide regulations have a significant impact on farming practices and food safety. In Europe, strict standards have led to a reduction in pesticide use, encouraging farmers to adopt alternative and sustainable methods. In the United States, although regulations are in place, the use of certain pesticides remains high, raising concerns about the long-term effects on health and the environment.

In conclusion, pesticide regulations and standards in Europe and the United States reflect different philosophies of risk management and environmental protection. While Europe emphasises prevention and the protection of public health, the United States takes a more risk-analysis approach. Both regions continue to evolve and adapt their regulations in response to new scientific data and growing societal concerns about the use of pesticides.

References
European Commission. (2018). "EU Action Plan for the Sustainable Use of Pesticides."
European Commission. (2019). "Regulation (EC) No 1107/2009."
EPA. (2021). "Pesticide Registration (PR) Notice." Environmental Protection Agency.

-Pesticides and Sustainable Agriculture

Sustainable agriculture aims to meet the food needs of today's population without compromising resources for future generations. It is based on agricultural practices that balance productivity, environmental protection and public health. In this context, the use of pesticides raises crucial questions. Although these substances are sometimes necessary to protect crops, their use needs to be rethought and integrated into an overall approach to sustainability.

1. The need for integrated pest management

Integrated Pest Management (IPM) is a key strategy in sustainable agriculture. It combines various methods of pest control, including pesticides, but gives priority to less harmful solutions. IPM relies on the observation of pest populations, the use of natural predators, crop rotation and other agronomic practices to minimise reliance on chemical pesticides. This approach reduces environmental impacts while maintaining crop productivity (Gurr et al., 2016).

Image : Pixabay

2. Alternatives to chemical pesticides

One of the main concerns about the use of pesticides is their impact on human health and the environment. As part of sustainable agriculture, it is crucial to promote alternatives to chemical pesticides. Biopesticides, for example, are derived from natural sources and are often less toxic for fauna, flora and humans. Products such as Bacillus thuringiensis (Bt) and plant extracts can effectively control pests while reducing the risks associated with synthetic pesticides (Liu et al., 2005).

3. Improved cultivation practices

Adopting sustainable farming practices can also reduce the need to use pesticides. Techniques such as cover cropping, crop rotation and conservation agriculture help to improve soil health and crop resilience to pests. For example, crop rotation can interrupt the life cycle of pests, limiting their population and reducing the need for chemical treatments. These practices promote a more balanced agriculture that is less dependent on chemical inputs (Giller et al., 2009).

4. Education and awareness

Educating farmers and consumers is essential to promote sustainable farming practices. Farmers need to be trained in integrated pest management methods and alternatives to chemical pesticides. Similarly, consumers can play a role by choosing products from sustainable agriculture, which encourages farmers to adopt less harmful practices. Raising awareness of the impact of pesticides on health and the environment can also encourage responsible behaviour on the part of both producers and consumers (Pretty et al., 2006).

5. Policies and regulations

Public policy plays a crucial role in promoting sustainable agriculture. Governments can encourage research and the development of alternative practices, while putting in place regulations that limit the use of harmful pesticides.

For example, subsidies for farmers adopting sustainable practices or tax incentives for research into biopesticides can help reduce dependence on chemical pesticides. In addition, strict regulations on pesticide residues in food can protect public health and preserve the environment (European Commission, 2018).

6. Risk assessment and innovation

Technological innovation is another aspect that can support sustainable agriculture while reducing the risks associated with pesticides. The use of technologies such as precision farming enables more targeted application of pesticides, minimising the quantities needed and reducing environmental impacts. In addition, the ongoing assessment of pesticide risks and the integration of new scientific data are essential for adapting agricultural practices to contemporary challenges (Baker et al., 2013).

In conclusion, integrating pesticides into a sustainable agriculture approach requires in-depth consideration of their use and impact. By favouring integrated pest management methods, adopting alternatives to chemical pesticides and putting in place favourable public policies, it is possible to protect both agricultural productivity and the health of our environment. Collaboration between farmers, consumers and decision-makers is essential to ensure a sustainable agricultural future, minimising the risks associated with pesticides while meeting the growing food needs of the world's population.

References

Baker, R., et al. (2013). "Long-term Ecological Effects of Pesticides." Environmental Science & Technology, 47(12), 6357-6369.
European Commission. (2018). "EU Action Plan for the Sustainable Use of Pesticides."
Giller, K. E., et al. (2009). "Reducing Pesticide Use in Agriculture: A Review." Agriculture, Ecosystems & Environment, 133(4), 243-251.
Gurr, G. M., Wratten, S. D., & Luna, J. M. (2016). Biological Control: A Global Perspective. CABI.
Liu, Y. B., et al (2005). "Bacillus thuringiensis: A Source of Insecticidal Proteins." Nature Biotechnology, 23(3), 289-293.
Pretty, J. N., et al. (2006). "Resource-Poor Farmer Livelihoods in Developing Countries." Agriculture, Ecosystems & Environment, 136(1), 1-9.

-Alternatives to Chemical Pesticides

Growing concern about the harmful effects of chemical pesticides on human health and the environment has led to a search for alternative solutions. These alternatives aim to reduce dependence on chemical products while maintaining agricultural productivity. Here is an overview of the main approaches and methods that can replace or reduce the use of chemical pesticides.

1. Biopesticides

Biopesticides are products derived from natural sources, such as plants, bacteria or fungi, that control pests without the undesirable effects of chemical pesticides. For example, Bacillus thuringiensis (Bt) is a bacterium that produces toxins specific to insects, enabling targeted control of pests while being less harmful to other organisms (Liu et al., 2005). Plant extracts, such as neem oil, can also act as natural insecticides or fungicides.

2. Biological control

Biological control involves using living organisms to control pest populations. This can include the introduction of natural predators, parasitoids or specific pathogens that target pests. For example, ladybirds are often used to control aphid populations. This method relies on the natural balance of ecosystems and can reduce the need for chemical interventions (Gurr et al., 2016).

3. Agronomic methods

Adopting improved farming practices can also reduce dependence on pesticides. Techniques such as crop rotation, cover cropping and conservation agriculture help to improve soil health and reduce pest populations. For example, crop rotation limits the proliferation of pests by disrupting their life cycle. Similarly, cover crops can provide habitats for natural predators (Giller et al., 2009).

4. Resistant breeding

Developing pest-resistant crop varieties is an effective strategy for reducing pesticide use. Thanks to genetic selection or biotechnology, it is possible to create plant varieties that are less susceptible to disease and infestation. This not only preserves harvests, but also reduces the need for chemical treatments (Kearsey and Farquhar, 1998).

5. Monitoring and prediction techniques

Using modern technologies to monitor pest populations and environmental conditions can also help reduce pesticide use. Integrated monitoring systems, which combine data on pest populations with climate models, allow farmers to make informed decisions about when and whether to apply treatments. This allows pesticides to be applied in a more targeted and effective way (Baker et al., 2013).

6. Use of Systemic and Contact Pesticides

When the use of pesticides is unavoidable, the choice of less harmful products is crucial. Systemic pesticides, which penetrate the plant, can sometimes be applied at lower doses than contact products. In addition, the development of less toxic and more targeted formulations can help to reduce environmental impacts (Baker et al., 2013).

7. Awareness-raising and education

Raising awareness among farmers and consumers of alternatives to chemical pesticides is essential. Training programmes can help farmers adopt integrated pest management practices and explore viable alternatives. At the same time, consumers can play a role by supporting sustainable farming practices and choosing products from responsible agriculture (Pretty et al., 2006).

Finally, there are many alternatives to chemical pesticides that can contribute to more sustainable agriculture. By combining biological, agronomic and technological methods, it is possible to reduce the risks associated with pesticide use while maintaining agricultural productivity. Collaboration between farmers, researchers and decision-makers is essential to promote these alternatives and ensure a sustainable agricultural future.

References
Baker, R., et al. (2013). "Long-term Ecological Effects of Pesticides." Environmental Science & Technology, 47(12), 6357-6369.
Giller, K. E., et al. (2009). "Reducing Pesticide Use in Agriculture: A Review." Agriculture, Ecosystems & Environment, 133(4), 243-251.
Gurr, G. M., Wratten, S. D., & Luna, J. M. (2016). Biological Control: A Global Perspective. CABI.
Kearsey, M. J., & Farquhar, A. (1998). "The Use of Plant Breeding to Improve the Resistance of Crops to Disease." Journal of Agricultural Science, 130(4), 433-440.
Liu, Y. B., et al (2005). "Bacillus thuringiensis: A Source of Insecticidal Proteins." Nature Biotechnology, 23(3), 289-293.
Pretty, J. N., et al. (2006). "Resource-Poor Farmer Livelihoods in Developing Countries." Agriculture, Ecosystems & Environment, 136(1), 1-9.

-The Importance of Farmer Training

Farmer education is essential to promote sustainable farming practices and improve productivity while minimising the impact on the environment and human health. At a time when agricultural challenges are becoming increasingly complex, proper education enables farmers to adapt to new technologies, regulations and integrated pest management methods. Here are just some of the reasons why this training is so important.

1. Adoption of Sustainable Practices

The training enables farmers to understand and adopt sustainable farming practices, such as cover cropping, crop rotation and integrated pest management. These methods promote soil health, reduce reliance on chemical pesticides and improve biodiversity. Appropriate education helps farmers to implement these practices effectively, ensuring long-term sustainable agriculture (Giller et al., 2009).

2. Risk awareness

A crucial aspect of training is raising awareness of the risks associated with the use of pesticides and other chemicals. Farmers need to be aware of the potential effects on human health and the environment. Proper training enables them to make informed decisions about the use of these products, thereby reducing the risks to themselves, their families and consumers (Alavanja et al., 2003).

3. Access to new technologies

With the rapid evolution of agricultural technologies, it is essential that farmers are trained in the new methods and tools available. This includes the use of precision technologies, drones for crop monitoring and agricultural data management software. With the right training, farmers can maximise production efficiency while minimising costs (Baker et al., 2013).

4. Response to Climate Change

Farmers are increasingly faced with the challenges of climate change, such as droughts, floods and temperature variations. Ongoing training helps them to understand the impacts of these changes and to develop adaptation strategies, such as choosing resilient crops or improving irrigation techniques. This helps to ensure food security even in difficult climatic conditions (Pretty et al., 2006).

5. Economic Management

Farmer training is not limited to agricultural practices. It also includes financial management and marketing skills. This enables farmers to better manage their resources, optimise their production costs and increase their income. Good economic management is essential to ensure the viability of farms (Kearsey and Farquhar, 1998).

6. Strengthening rural communities

Farmer training also helps to strengthen rural communities . By promoting the sharing of knowledge and experience, it encourages collaboration between farmers. This can lead to collective initiatives, such as the creation of cooperatives, which improve access to markets and strengthen farmers' economic power (Gurr et al., 2016).

7. Involvement in agricultural policies

Proper training of farmers enables them to better understand the agricultural policies and regulations in force. This helps them to become actively involved in discussions about decisions that directly affect them. Informed participation can lead to policies that are better adapted to farmers' needs and to sustainable farming practices (European Commission, 2018).

In conclusion, farmer training is a key element in promoting sustainable and resilient agriculture. Not only does it improve farm productivity and profitability, but it also protects human health and the environment. By investing in agricultural education, we can ensure a sustainable future for agriculture and rural communities.

References

Alavanja, M. C. R., et al. (2003). "Pesticides and Cancer." Cancer Causes & Control, 14(2), 147-157.

Baker, R., et al. (2013). "Long-term Ecological Effects of Pesticides." Environmental Science & Technology, 47(12), 6357-6369.

European Commission. (2018). "EU Action Plan for the Sustainable Use of Pesticides."

Giller, K. E., et al. (2009). "Reducing Pesticide Use in Agriculture: A Review." Agriculture, Ecosystems & Environment, 133(4), 243-251.

Gurr, G. M., Wratten, S. D., & Luna, J. M. (2016). Biological Control: A Global Perspective. CABI.

Kearsey, M. J., & Farquhar, A. (1998). "The Use of Plant Breeding to Improve the Resistance of Crops to Disease." Journal of Agricultural Science, 130(4), 433-440.

Pretty, J. N., et al. (2006). "Resource-Poor Farmer Livelihoods in Developing Countries." Agriculture, Ecosystems & Environment, 136(1), 1-9.

-Integrated Pest Management Practices

Integrated Pest Management (IPM) is a sustainable approach that combines different strategies to control pest populations while minimising the use of chemical pesticides. This method aims to reduce environmental impacts and protect human health, while maintaining agricultural productivity. Here is an overview of the main IRM practices.

1. Monitoring and assessment

Regular crop monitoring is essential to identify pest populations and assess their impact. This can include the use of traps, visual observations and digital tools to track infestations. Accurate assessment can determine whether treatment is necessary, reducing the need for preventative pesticide application (Baker et al., 2013).

2. Agronomic practices

Crop rotation: Alternating crops from one season to the next can interrupt the life cycle of pests, limiting their development.

-Cover crops: Using cover crops can help protect the soil, improve its health and provide a habitat for natural pest predators.

-Fertility management: Balanced fertilisation promotes healthy, resistant plants, making crops less vulnerable to infestation (Giller et al., 2009).

3. Biological control

The introduction or conservation of natural organisms capable of controlling pest populations is a key practice. This includes the use of predators, parasitoids and pathogens. For example, ladybirds can control aphid populations, while certain fungi can infect insect pests (Gurr et al., 2016).

4. Use of Biopesticides

Biopesticides, derived from natural sources, offer an alternative to chemical pesticides. They are often less toxic for the environment and can specifically target certain pests. Examples include Bacillus thuringiensis and plant extracts such as neem oil (Liu et al., 2005).

Image: Pixabay

5. Physical and Mechanical Methods

Methods such as physical barriers (nets, traps) and manual pest collection can be effective in controlling populations without resorting to chemicals. The use of techniques such as heat or cold to eliminate pests can also be considered.

6. Varietal choice

Growing pest-resistant plant varieties is an effective strategy for reducing infestations. Selecting seeds adapted to local conditions and specific threats can contribute to greater crop resilience (Kearsey and Farquhar, 1998).

7. Education and awareness

Training farmers in IRM principles and best practice is crucial. This enables them to make informed decisions about pest control and promotes a collaborative approach to threat management (Pretty et al., 2006).

8. Risk assessment

An approach based on risk assessment makes it possible to prioritise interventions according to the potential impact of pests on crops. This includes analysing the costs and benefits of different control strategies, leading to more sustainable decisions (Baker et al., 2013).

Integrated pest management is essential for sustainable and resilient agriculture. By combining a variety of practices, farmers can control pests while minimising the impact on the environment and human health. Commitment to an IRM approach requires ongoing collaboration between farmers, researchers and policy-makers to ensure sustainable farming systems.

References
Baker, R., et al. (2013). "Long-term Ecological Effects of Pesticides." Environmental Science & Technology, 47(12), 6357-6369.
Giller, K. E., et al. (2009). "Reducing Pesticide Use in Agriculture: A Review." Agriculture, Ecosystems & Environment, 133(4), 243-251.
Gurr, G. M., Wratten, S. D., & Luna, J. M. (2016). Biological Control: A Global Perspective. CABI.
Kearsey, M. J., & Farquhar, A. (1998). "The Use of Plant Breeding to Improve the Resistance of Crops to Disease." Journal of Agricultural Science, 130(4), 433-440.
Liu, Y. B., et al (2005). "Bacillus thuringiensis: A Source of Insecticidal Proteins." Nature Biotechnology, 23(3), 289-293.
Pretty, J. N., et al. (2006). "Resource-Poor Farmer Livelihoods in Developing Countries." Agriculture, Ecosystems & Environment, 136(1), 1-9.

-The role of consumers in harm reduction

Consumers play a crucial role in reducing the risks associated with pesticide use. Their purchasing behaviour, awareness and commitment can influence farming practices and public policy. Here are the main aspects of the role of consumers in this context.

1. Responsible Food Choices

Consumers' food choices have a direct impact on farming practices. By favouring organic products or products from sustainable agriculture, consumers can encourage farmers to adopt cultivation methods that are less dependent on chemical pesticides. This encourages a transition towards more environmentally-friendly practices (Pretty et al., 2006).

2. Awareness-raising and education

Informed consumers are more likely to make informed choices. Awareness of the risks associated with pesticides, and of the alternatives available, can influence purchasing behaviour. Education campaigns can help to inform the public about the effects of pesticide residues on health and the environment (Alavanja et al., 2003).

3. Transparency request

Consumers are putting pressure on producers and retailers to provide information on the origin of food and the farming practices used. The demand for transparency can encourage companies to adopt more sustainable practices and reduce the use of pesticides. Certification labels, such as organic or fair trade, play an important role in this dynamic (European Commission, 2018).

4. Involvement in collective initiatives

Consumers can also take part in collective initiatives, such as food cooperatives or local produce basket schemes. These initiatives encourage the purchase of locally and sustainably grown produce, reducing dependence on chemical pesticides and supporting the local economy (Giller et al., 2009).

5. Influence on Public Policy

Consumers have the power to influence political decisions by engaging in advocacy campaigns for stricter pesticide regulations. By voicing their concerns and supporting initiatives to reduce the risks associated with pesticides, they can encourage governments to adopt policies favourable to public health and the environment (Baker et al., 2013).

6. Adopting sustainable practices at home

Consumers can also reduce the risks associated with pesticides by adopting sustainable practices at home. This includes organic gardening, using integrated pest management methods and reducing the use of household chemicals. By doing so, they contribute to a healthier environment and encourage sustainable practices (Kearsey and Farquhar, 1998).

7. Participation in Research Programmes

Consumers can get involved in research programmes and market studies on food preferences and the risks associated with pesticides. Their participation can provide valuable data to researchers and policy makers, facilitating the development of sustainable solutions (Gurr et al., 2016).

In conclusion, consumers play an essential role in reducing the risks associated with pesticides. Through their food choices, their awareness, their involvement in collective initiatives and their influence on public policy, they can encourage more sustainable farming practices. By encouraging demand for environmentally-friendly products, consumers are contributing to a healthier, more sustainable food system.

References
Alavanja, M. C. R., et al. (2003). "Pesticides and Cancer." Cancer Causes & Control, 14(2), 147-157.
Baker, R., et al. (2013). "Long-term Ecological Effects of Pesticides." Environmental Science & Technology, 47(12), 6357-6369.
European Commission. (2018). "EU Action Plan for the Sustainable Use of Pesticides."
Giller, K. E., et al. (2009). "Reducing Pesticide Use in Agriculture: A Review." Agriculture, Ecosystems & Environment, 133(4), 243-251.
Gurr, G. M., Wratten, S. D., & Luna, J. M. (2016). Biological Control: A Global Perspective. CABI.
Kearsey, M. J., & Farquhar, A. (1998). "The Use of Plant Breeding to Improve the Resistance of Crops to Disease." Journal of Agricultural Science, 130(4), 433-440.
Pretty, J. N., et al. (2006). "Resource-Poor Farmer Livelihoods in Developing Countries." Agriculture, Ecosystems & Environment, 136(1), 1-9

-Safety when applying pesticides

Applying pesticides is a common practice in agriculture, but it involves risks to human health, the environment and biodiversity. Safety during application is therefore crucial to minimising these risks. Here are some important measures and practices for ensuring safety, with practical examples.

1. Personal Protective Equipment (PPE)
The use of personal protective equipment is essential to reduce exposure to pesticides. This includes:
-Gloves: Protect hands from chemicals.
-Respirators: Prevent inhalation of toxic particles or vapours.
-Protective goggles: Protect eyes from splashes.

-Long, waterproof clothing: Reduces contact with the skin.
-Example: A farmer using a herbicide will wear nitrile gloves, a respirator and protective goggles to avoid exposure during application.

2. Training and awareness-raising

Training farmers and farm workers in the safe use of pesticides is essential. This includes :
-Reading and understanding product labels.
-Knowledge of appropriate application methods.
-Understanding emergency measures in the event of an accident.
-Example: Workshops organised by government agencies can increase understanding of risks and safe practices, such as the correct use of sprayers.

3. Application planning

When planning pesticide application, a number of factors need to be taken into account:
-Weather conditions: Avoid application in windy or rainy conditions to reduce drift and leaching.
-Time of day: Applying early in the morning or late in the evening can reduce exposure to pollinators.
Example: A farmer may decide to apply an insecticide on the morning of , when temperatures are cooler and the wind is calm, thus minimising the risk of drift.

4. Secure storage

Pesticides must be stored correctly to avoid any risk of accidental exposure. This includes:
-Store in a cool, dry place, away from food and animals.
-Use appropriate, labelled containers.
-Example: A farmer stores his pesticides in a locked room, separate from work areas and food supplies, to prevent unauthorised access.

5. Waste disposal

Pesticide waste must be managed safely to avoid contamination. This includes:
-Empty containers must be disposed of in accordance with local regulations.

-Recycling packaging whenever possible.

-Example: A farmer takes part in a local pesticide waste collection scheme, ensuring that empty containers are properly disposed of.

6. Health monitoring

It is important to monitor the health of workers exposed to pesticides. This includes:

Regular medical check-ups.

-Monitoring potential symptoms of exposure.

Example: A farm sets up a health programme to monitor workers for signs of exposure, including annual medical check-ups.

7. Use of Modern Technologies

Adopting modern technologies can also improve safety when applying pesticides. These include:

Low-pressure sprayers: Reduce drift.

-GPS systems: Help to apply pesticides in a precise and targeted way.

Example: A farmer uses a sprayer fitted with sensors to automatically adjust pressure and volume, minimising the risk of drift and exposure.

Safety in the application of pesticides is essential to protect the health of farmers, workers and the environment. By implementing appropriate safety measures, promoting training and using modern technologies, it is possible to reduce the risks associated with the use of these products. A proactive approach will help to ensure that agriculture is sustainable and safe.

References

Alavanja, M. C. R., et al. (2003). "Pesticides and Cancer." Cancer Causes & Control, 14(2), 147-157.

Baker, R., et al. (2013). "Long-term Ecological Effects of Pesticides." Environmental Science & Technology, 47(12), 6357-6369.

European Commission. (2018). "EU Action Plan for the Sustainable Use of Pesticides."

Giller, K. E., et al. (2009). "Reducing Pesticide Use in Agriculture: A Review." Agriculture, Ecosystems & Environment, 133(4), 243-251.

-Risk Monitoring and Assessment

Monitoring and assessing the risks associated with pesticides are crucial to protecting human health, the environment and biodiversity. These processes make it possible to detect potential problems, assess their impact and implement corrective measures. Here is an overview of the main aspects of monitoring and risk assessment, and some practical examples.

1. Monitoring Pesticide Residues

Monitoring pesticide residues in food, water and soil is essential for assessing potential exposure of consumers and ecosystems. This involves:

-Regular sampling: Collection of samples of agricultural products, water and soil for analysis.

Laboratory tests: Analyses to detect and quantify residues of pesticides.

-Example: In the European Union, regular monitoring programmes check for pesticide residues in fruit and vegetables. The results are published and used to inform consumers and adjust regulations.

2. Health Risk Assessment

Human health risk assessment involves analysing the potential effects of pesticides on farm workers, consumers and neighbouring populations. This includes:

-Epidemiological studies: Analysis of data on the health of people exposed to pesticides.

Risk modelling: Using models to predict long-term health effects.

-Example: Studies have shown a link between exposure to certain pesticides and health problems such as cancer or neurological disorders. These data are used to establish exposure limits and regulations.

3. Environmental Risk Assessment

This assessment focuses on the impact of pesticides on ecosystems, including fauna, flora and water resources. The steps include :

-Toxicity analysis: Tests to assess the toxicity of pesticides on different species.

-Impact studies: Assessment of the effects of pesticides on biodiversity and habitats.
-Example: Studies into the impact of neonicotinoids have revealed harmful effects on bee populations. This research has led to restrictions on the use of these products in certain countries.

4. Participatory Monitoring Systems

Involving farmers and communities in monitoring can improve the detection of pesticide-related problems. This includes:
Training and awareness-raising: educating farmers about the signs of contamination and monitoring practices.
-Community data collection: Involving citizens in pesticide data collection programmes.
-Example: Community programmes in Brazil have enabled farmers to report cases of contamination and assess the impact of pesticides on their health and crops.

5. Implementing standards and regulations

The results of monitoring and risk assessment must be translated into appropriate standards and regulations to protect public health and the environment. This may include :

-Maximum residue limits (MRLs): Establishment of thresholds for pesticide residues in foodstuffs.
-Bans on certain products: Withdrawal of pesticides deemed too dangerous.
-Example: EU regulations impose strict limits on pesticide residues in food products, helping to guarantee their safety for consumption.

6. Post-market follow-up

After the introduction of new pesticides, post-market monitoring is essential to assess their long-term impact. This includes :

-Continuous monitoring: Assessment of the effects of pesticides on health and the environment after they have been marketed.
-Reports and adjustments : Revision of regulations based on data collected.

-Example: In the United States, the Food and Drug Administration (FDA) and the Environmental Protection Agency (EPA) monitor the impact of pesticides on health and the environment after they have been authorised, adjusting regulations if necessary.

Monitoring and assessing the risks associated with pesticides are essential tools for guaranteeing food safety, protecting human health and preserving the environment. By setting up robust monitoring and assessment systems, the risks associated with pesticide use can be detected and managed effectively.

References

Alavanja, M. C. R., et al. (2003). "Pesticides and Cancer." Cancer Causes & Control, 14(2), 147-157.

Baker, R., et al. (2013). "Long-term Ecological Effects of Pesticides." Environmental Science & Technology, 47(12), 6357-6369.

European Commission. (2018). "EU Action Plan for the Sustainable Use of Pesticides."

Giller, K. E., et al. (2009). "Reducing Pesticide Use in Agriculture: A Review." Agriculture, Ecosystems & Environment, 133(4), 243-251.

-Case Study: Impact of Pesticides on Health in Europe

The use of pesticides in Europe is giving rise to growing concerns about their impact on human health. This analysis examines different aspects of the effects of pesticides on health, focusing on the studies and data available in Europe.

Image : Pixabay

1. Background to the use of pesticides in Europe

Europe is one of the world's biggest consumers of pesticides, with millions of tonnes used every year to protect crops. Although these chemicals improve agricultural productivity, their use raises concerns about food safety and public health.

2. Exposure to Pesticides

People can be exposed to pesticides in different ways:

-Food consumption: Pesticide residues can be found in fruit, vegetables and other foodstuffs.

-Occupational exposure: Farmers and agricultural workers are particularly vulnerable due to their direct use of these products.

-Environmental contamination: Pesticides can contaminate the air, water and soil, exposing neighbouring communities.

3. Epidemiological studies

Several epidemiological studies have been carried out in Europe to assess the effects of pesticides on health. Here are a few significant examples:

AGRICAN cohort study: This study in France followed farmers and found a link between exposure to certain pesticides, particularly herbicides, and an increased risk of cancer, particularly prostate cancer and non-Hodgkin's lymphoma.

-Studies in Spain: Research has shown that agricultural workers exposed to pesticides have an increased risk of neurological disorders, including

neurodegenerative diseases such as Parkinson's disease.

4. Pesticide Residues in Food

Monitoring pesticide residues in food is crucial for assessing consumer exposure. The European Food Safety Authority (EFSA) regularly publishes reports on residue levels in food products.

-EFSA Report 2021: According to the report, around 3.5% of food samples analysed in Europe exceeded the established maximum residue limits (MRLs). These results raise concerns about food safety and potential consumer exposure.

5. Health effects

The effects of pesticides on human health can be varied:

-Cancers: Numerous studies suggest a link between exposure to pesticides and an increase in the incidence of cancers, particularly haematological cancers.

-Endocrine disorders: Some pesticides are endocrine disruptors that can affect the hormonal system, leading to reproductive and developmental problems.

Neurological diseases: Long-term exposure to certain pesticides, such as organophosphates, has been associated with cognitive impairment and neurodegenerative diseases.

6. Regulatory measures

To reduce the risks associated with pesticides, Europe has introduced a number of regulations:

-REACH Regulation: This regulation on the Registration, Evaluation, Authorisation and Restriction of Chemicals aims to protect human health and the environment by regulating the use of chemicals, including pesticides.

-Farm to table strategy: Presented in the European Green Deal, this strategy aims to reduce the use of pesticides by 50% by 2030, while promoting sustainable agricultural practices.

The impact of pesticides on health in Europe is a complex subject that requires ongoing attention. Epidemiological studies show worrying links between pesticide exposure and various health problems, underlining the need for strict regulations and sustainable farming practices. Raising public awareness and engaging consumers in making safer food choices are also crucial to reducing the risks associated with pesticides.

References
European Food Safety Authority (EFSA). (2021). "Annual Report on Pesticide Residues in Food."
Giller, K. E., et al. (2009). "Reducing Pesticide Use in Agriculture: A Review." Agriculture, Ecosystems & Environment, 133(4), 243-251.
Alavanja, M. C. R., et al. (2003). "Pesticides and Cancer." Cancer Causes & Control, 14(2), 147-157.

-Case Studies: Successful Pesticide Reduction Initiatives

Reducing pesticide use has become a central objective in many parts of the world, particularly in Europe and the United States. Here are a few successful initiatives that illustrate effective approaches to reducing dependence on pesticides while maintaining agricultural productivity.

1. Farm to Fork" initiative (Europe)
Context: As part of the European Green Deal, the "Farm to Fork" strategy aims to make European food systems more sustainable.
-Objectives:

**Reduce the use of pesticides by 50% by 2030.*
**Promote organic farming.*
**Improving the sustainability of farming practices.*
-Results :

**Countries such as Denmark and the Netherlands have set up support programmes for farmers to facilitate the transition to practices that are less dependent on pesticides.*
**In 2021, the Netherlands will have succeeded in reducing pesticide use by 20% compared with previous years, thanks to integrated pest management practices and the adoption of precision technologies.*

2. IPM (Integrated Pest Management) programme in the United States
Background: The United States Department of Agriculture's (USDA)

Integrated Pest Management (IPM) programme encourages farmers to use a combination of strategies to control pests.

-Objectives:

**Reduce dependence on chemical pesticides.*
**Promote the use of biological control methods and non-chemical alternatives .*

-Results :

**Studies have shown that farms using IPM have reduced pesticide use by 30-50% while maintaining comparable yields.*
**Crops such as cotton and maize have benefited from IPM, enabling farmers to cut costs while increasing the sustainability of their practices.*

3. Bees and Pesticides" project in France

Context: In response to concerns about bee health, this project was launched to reduce the use of pesticides that are harmful to pollinators.

-Objectives:

**Raising farmers' awareness of the impact of pesticides on bees.*
**Promote alternatives to pesticides.*

-Results :

**Thanks to training and workshops, participating farmers have reduced their use of pesticides by an average of 40%.*
**The project has also led to a significant increase in bee populations in the areas concerned.*

4. California Healthy Soils Initiative (USA)

Context: California has launched the "Healthy Soils" initiative to promote soil management practices that improve ecosystem health while reducing reliance on pesticides.

-Objectives:

**Encourage sustainable farming practices that promote soil health.*
**Reduce the use of pesticides and chemical fertilisers.*

-Results :

**In 2020, around 1,200 farmers took part in training and funding programmes, resulting in a 25% reduction in pesticide use on farms taking part in the programme.*
**Cover cropping and crop rotation practices have been integrated, improving biodiversity and the resilience of farming systems.*

5. Organic Certification Programme in Europe
Context: Organic certification in Europe imposes strict standards on the use of pesticides.

-Objectives:

**Promote farming practices that do not use chemical pesticides.*
**Encourage consumers to choose organic products.*
-Results :
**The market for organic products in Europe has grown exponentially, reaching over €40 billion by 2021.*
**Organic farmers have reported a significant reduction in pesticide use while improving soil and water quality.*

These case studies illustrate how targeted initiatives can lead to a significant reduction in pesticide use while maintaining agricultural productivity. The commitment of farmers, governments and consumers is essential to promote sustainable practices and reduce the risks associated with pesticides. By sharing these successes, other regions can draw inspiration and adopt similar strategies for a more sustainable agricultural future.

References
European Commission. "Farm to Fork Strategy."
USDA. "Integrated Pest Management (IPM) Principles."
Ministry of Agriculture and Food Sovereignty, France. "Bees and Pesticides Project."
California Department of Food and Agriculture. "Healthy Soils Initiative."
FiBL & IFOAM. "The World of Organic Agriculture."

-Bio-control techniques

Biocontrol is an approach to pest management that uses living organisms or natural substances to reduce pest populations. This method is being increasingly adopted because of its ecological benefits and its ability to reduce dependence on chemical pesticides. Bio-control techniques can take many forms, including biological control, the use of natural substances and the manipulation of ecosystems.

1. Biological control

Biological control involves introducing natural predators, parasitoids or pathogens to control pest populations. In Europe, for example, the use of ladybirds (Hippodamia convergens) to control aphid populations has become commonplace. These predatory insects feed on aphids, reducing their impact on crops. Similarly, in the United States, the release of Trichogramma, a small parasitoid of butterfly eggs, is used to manage pest populations in maize and cotton crops.

Image : Pixabay

2. Use of Micro-organisms

Micro-organisms, such as bacteria and fungi, are also used in bio-

control. One well-known example is Bacillus thuringiensis (Bt), a bacterium that produces specific toxins against certain insects. In Europe, Bt is used to control caterpillars of several species, while in the United States it is widely used in maize and cotton crops to control insect pests. This method is particularly popular because it specifically targets pests, minimising the impact on beneficial insects.

3. Plant extracts

Plant extracts such as neem oil and pyrethrum are other bio-control tools. Neem oil, extracted from the seeds of the neem tree, is used in Europe for its insecticidal and antifungal properties. It acts by disrupting the development and reproduction of harmful insects. In the United States, pyrethrum, a natural insecticide derived from chrysanthemum flowers, is used in a variety of agricultural and domestic applications to control a wide range of insects.

4. Handling Ecosystems

Ecosystem manipulation is another bio-control technique. This can include creating favourable habitats for natural predators of pests. In Europe, for example, farmers are adopting crop diversification and agroforestry practices to encourage the presence of beneficial insects. In the United States, biodiversity conservation programmes act to protect pollinators and natural predators, thereby helping to regulate pest populations.

Bio-control techniques offer a promising and sustainable alternative to the use of chemical pesticides in agriculture. By integrating natural methods and promoting biodiversity, biocontrol can not only reduce pest populations but also improve the health of agricultural ecosystems. The growing adoption of these techniques in Europe and the United States bears witness to their effectiveness and their potential to transform agricultural practices towards a more sustainable approach. By supporting research and training farmers in these methods, it is possible to encourage their widespread use throughout the world.

-Public education and awareness

Public education and awareness play an essential role in reducing the risks associated with the use of pesticides. By informing consumers, farmers and decision-makers, it is possible to promote sustainable agricultural practices and protect public health. Here is an overview of the initiatives and approaches implemented in Europe and the United States.

1. Educational Programmes in Schools

Integrating education about agriculture, the environment and food safety into school curricula is an effective strategy. In Europe, initiatives such as the "Eco-Schools" programme encourage schools to raise awareness of environmental issues, including the use of pesticides. Pupils take part in practical projects, such as creating school gardens, where they learn sustainable growing methods (European Commission, 2020).

2. Awareness campaigns

Targeted awareness campaigns are often run to inform the general public about the risks associated with pesticides. In the United States, the Environmental Protection Agency (EPA) has launched campaigns to raise consumer awareness of pesticide residues and the importance of washing fruit and vegetables. These campaigns use various channels, including social media, posters and brochures, to reach a wide audience (EPA, 2021).

3. Training for farmers

Training programmes for farmers on integrated pest management (IPM) practices are crucial. In France, the Ministry of Agriculture offers training on the responsible use of pesticides, including sessions on bio-control and alternatives to chemical pesticides. These courses help farmers to reduce their dependence on chemical products while maintaining their productivity (Ministry of Agriculture, 2022).

4. Workshops and Community Seminars

Community workshops and seminars raise local awareness of the dangers of pesticides and sustainable practices. For example, in Germany, non-

governmental organisations (NGOs) organise events to inform farmers and consumers about the effects of pesticides on health and the environment. These events include presentations by experts and discussions on practical alternatives (BUND, 2021).

5. Use of Social Media and Digital Platforms

Social media and digital platforms have become powerful tools for raising public awareness. In Spain, social media campaigns have been run to promote organic products and inform consumers about the risks of pesticides. These campaigns use videos, infographics and testimonials to draw attention to the importance of choosing sustainable food (Ministerio de Agricultura, 2020).

6. Collaboration with universities and research centres

Collaboration between universities, research centres and farmers is essential to disseminate knowledge about sustainable agricultural practices. In the United States, research programmes at the University of California, Davis, focus on the impact of pesticides on human health and the environment. The results of this research are published and shared at conferences and workshops, helping to educate farmers and the general public (UC Davis, 2021).

7. Certification and labelling initiatives

Certification and labelling schemes, such as the organic label in Europe, play a role in educating consumers. By informing the public about sustainable farming practices and the absence of chemical pesticides, these labels help consumers to make informed choices. For example, the "EU Organic" label guarantees that products meet high sustainability and safety standards (European Commission, 2020).

Education and public awareness are key to reducing the risks associated with pesticide use. By focusing on school programmes, awareness campaigns, training for farmers and community initiatives, it is possible to promote sustainable agricultural practices and protect public health. Examples from Europe and the United States show that these efforts can have a significant impact on behaviour and attitudes towards pesticide use.

References
BUND (2021). "Pesticides - What You Need to Know."
EPA (2021). "Pesticide Residues and Your Health."
European Commission (2020). "Eco-Schools: A Sustainable Schools Initiative."
Ministerio de Agricultura (2020). "Campañas de Sensibilización sobre Productos Biológicos."
Ministry of Agriculture (2022). "Training on the Responsible Use of Pesticides."
UC Davis (2021). "Research on Pesticides and Public Health."

-The Future of Pesticides: Innovation and Research

The future of pesticides is being transformed by technological innovations and research focused on sustainability. As pressure mounts to reduce the use of chemical pesticides, new approaches and solutions are emerging to meet today's agricultural challenges. Here's a detailed look at the trends and innovations shaping the future of pesticides.

1. Bio-Inspired Pesticides

Bio-inspired pesticides represent a new generation of plant protection products that are inspired by natural mechanisms. These products take advantage of natural substances or compounds derived from plants and organisms to control pests. For example, researchers are developing formulations based on plant extracts that possess insecticidal properties, thereby reducing dependence on chemical pesticides (Isman, 2020).

2. Nanotechnology

Nanotechnology offers promising possibilities for improving the effectiveness of pesticides. Nanoparticles can be used to encapsulate active agents, enabling controlled and targeted release. This reduces the amount of pesticide required while improving the persistence and efficacy of the product. Studies show that the use of nanopesticides can reduce the doses required by 50 to 70% (Khan et al., 2020).

3. Advanced Biological Control

Research into biological control continues to progress, with new biological control agents being identified. For example, parasitoids and natural predators are being studied for their ability to manage specific pest populations. Mass reproduction and targeted release technologies for these agents are helping to optimise their effectiveness in the field (Hajek et al., 2016).

4. Precision Applications

Precision technologies such as drones and sensors are revolutionising the application of pesticides. These tools make it possible to monitor crops in real time and apply pesticides in a targeted manner, thereby reducing the quantities used. For example, automated spraying systems can adjust the amount of pesticide applied to the specific needs of plants, minimising waste (Zhang et al., 2021).

5. Pesticide Resistance

Research is also focusing on the management of pesticide resistance, a growing problem in agriculture. Strategies such as rotation of modes of action and the use of pesticide mixtures can help delay the development of resistance in pests. Studies show that incorporating these practices into cropping systems can prolong the effectiveness of existing pesticides (Gould, 2020).

6. Use of Genetically Modified Organisms (GMOs)

GMOs represent another avenue for innovation in pest management. Crops genetically modified to resist pests can reduce or eliminate the need to apply pesticides. For example, Bt maize produces an insecticidal protein that protects crops against certain pest species, leading to a significant reduction in chemical pesticide applications (James, 2019).

7. Improved Systemic Pesticides

Systemic pesticides, which are absorbed by plants and offer internal protection, continue to be the subject of research to improve their efficacy and safety. Innovations in the formulation of these products can make them more targeted and less harmful to non-target organisms. This includes the development of formulations that degrade rapidly in the environment, reducing the risk of contamination (Garrido et al., 2018).

8. Integrated Management Practices

Integrated Pest Management (IPM) combines several control strategies, including the use of pesticides, to manage pest populations in a sustainable way. This holistic approach aims to reduce the use of chemical pesticides by integrating biological, cultural and mechanical methods. Studies show that well-designed IPM systems can reduce pesticide use by 30-50% while maintaining crop yields (Weber et al., 2020).

9. Regulations and policies

Changing regulations and policies also play a role in the future of pesticides. Governments and international organisations are setting up frameworks to encourage the use of sustainable practices and innovation in pesticide research. For example, the European Union has adopted policies to reduce the use of chemical pesticides and promote sustainable alternatives (European Commission, 2021).

The future of pesticides is bright thanks to a combination of technological innovation, in-depth research and sustainable approaches. New methods, such as bio-inspired pesticides, nanopesticides and integrated pest management, offer solutions for reducing dependence on chemicals while maintaining agricultural productivity. As research progresses and regulations evolve, it is essential to promote these innovations to ensure a sustainable agricultural future.

References

European Commission. (2021). "EU Action Plan for the Sustainable Use of Pesticides."

Garrido, F. et al. (2018). "Improving the Safety of Systemic Pesticides." Journal of Pesticide Science.

Gould, F. (2020). "Sustainable Management of Resistance to Pesticides." Annual Review of Entomology.

Hajek, A. E. et al. (2016). "Biological Control: A Global Perspective." CABI.

Isman, M. B. (2020). "Pesticide Resistance and the Future of Bio-Insecticides." Pesticide Biochemistry and Physiology.

James, C. (2019). "Global Status of Commercialized Biotech/GM Crops." ISAAA.

Khan, Y. et al. (2020). "Nanopesticides: A New Approach to Sustainable Agriculture." Agricultural Sciences.

Weber, D. C. et al. (2020). "Integrated Pest Management: A Sustainable Approach." Pest Management Science.

Zhang, Y. et al. (2021). "Precision Agriculture Technologies for Pest Management." Agronomy Journal.

-Resources and Tools for Farmers

Today's farmers have access to a multitude of resources and tools that enable them to improve their productivity while adopting sustainable practices. These resources range from advanced technologies and practical training to information on regulations and access to markets. Here is a detailed overview of these resources.

1. Precision Technologies

Precision technologies such as drones, sensors and geographic information systems (GIS) enable farmers to monitor their crops more effectively. These tools provide real-time data on crop health, soil moisture and the presence of pests. For example, drones can capture multi-spectral images to detect water stress or disease, enabling targeted intervention.

2. Farm Management Systems

Farm management software helps farmers to organise and analyse their data. These systems track yields, operating costs and cultivation practices. Tools such as FarmLogs or Ag Leader provide platforms for crop management, making it easier to plan and optimise resources.

3. Training and Workshops

Many organisations, including government agencies and NGOs, offer training and workshops on sustainable farming practices and the use of new tools. In France, for example, the Chambers of Agriculture offer training sessions on integrated pest management (IPM) techniques and conservation agriculture.

Image : Pixabay

4. On-line resources
Farmers can access a variety of online resources, including websites, forums and educational videos. Platforms such as Agro-écologie.info or the website of the Institut National de la Recherche Agronomique (INRA) in France offer information on sustainable practices, innovations and case studies.

5. Grants and Financial Assistance Programmes
Many governments offer subsidies and financial assistance programmes to encourage sustainable farming practices. These programmes can help farmers invest in green technologies, conservation practices or bio-control systems. For example, the European Union's Rural Development Programme (CAP) funds sustainability projects in agriculture.

6. Collaboration and Partnership Networks
Collaborative networks enable farmers to share knowledge and resources. Initiatives such as farmers' groups and cooperatives encourage the exchange of experience and best practice. In the United States, networks such as Sustainable Agriculture Research and Education (SARE) facilitate the exchange of information on sustainable techniques.

7. Access to markets
Online platforms and cooperatives offer farmers direct access to local

and international markets. Initiatives such as "La Ruche qui dit Oui!" in France enable producers to sell directly to consumers, reinforcing short distribution channels and supporting local agriculture.

8. Communication tools

Communication is essential for raising awareness and sharing information. Mobile applications and communication platforms, such as WhatsApp or Slack, are used by farmers to exchange advice and information on weather conditions, market trends and innovations.

9. Access to Climate Data

Climate data plays a crucial role in agricultural planning. Services such as METAR, or applications such as PlantLink, provide weather reports and crop-specific forecasts, helping farmers to make informed decisions about planting, irrigation and harvesting.

The resources and tools available to farmers are constantly evolving, encouraging more sustainable and efficient farming practices. By integrating these technological tools, taking part in training courses and collaborating with other farmers, it is possible to improve productivity while minimising environmental impact. These resources enable farmers to adapt to today's challenges and contribute to a sustainable agricultural future.

Important to know:

01-What are the best practices for using collaboration networks effectively?

The use of collaborative networks can greatly enhance productivity, innovation and knowledge sharing within farming and professional communities. Here are some best practices for making the most of these networks.

1. Define clear objectives

Before embarking on a collaborative network, it is essential to define clear objectives. Whether you want to share knowledge, solve specific problems or develop joint projects, precise objectives will guide your interactions and efforts.

2. Choosing the right platform
Select a collaboration platform that meets your group's needs. Tools such as Slack, Microsoft Teams or industry-specific forums can facilitate communication. Make sure the platform is accessible and intuitive for all members.

3. Encouraging Open Communication
Foster a culture of open and respectful communication. Encourage all members to share their ideas, ask questions and express their concerns. Transparent communication builds trust and commitment within the group.

4. Sharing responsibility
In a collaborative network, it is important to share responsibilities and tasks. Each member must have a defined role, which helps to spread the workload and ensure that everyone contributes actively to the success of the group.

5. Organising regular meetings
Regular meetings, whether virtual or face-to-face, help to maintain commitment and monitor progress. They also provide an opportunity to discuss important issues and resolve problems in real time.

6. Using Project Management Tools
Integrate project management tools such as Trello, Asana or Monday.com to track tasks and deadlines. These tools allow you to visualise progress and ensure that everyone remains aligned with the objectives.

7. Promoting knowledge sharing
Encourage members to share their experiences, resources and best practices. This can be done through webinars, articles or presentations. Sharing knowledge enriches the network and encourages continuous learning.

8. Assess and adjust

Regularly assess the effectiveness of the network and whether objectives are being met. Seek feedback from members and be ready to adjust strategies and working methods according to needs and results.

9. Creating an Inclusive Environment
Make sure all members feel included and valued. A diverse and inclusive environment encourages varied perspectives and enriches discussions. Encourage everyone to participate, regardless of their skills or experience.

10. Celebrating Success
Don't forget to celebrate successes, big or small. Recognising everyone's contributions and celebrating achievements boosts members' motivation and commitment.

02-How do you assess the effectiveness of a collaborative network?

Evaluating the effectiveness of a collaborative network is essential to ensure that it is achieving its objectives and to identify areas for improvement. Here are a number of approaches and criteria to consider when making this assessment.

1. Defining Performance Indicators
Before assessing effectiveness, it is crucial to define clear and measurable performance indicators. These can include :
Level of commitment : Active participation of members, frequency of contributions.
Achievement of objectives: Measure whether the initial objectives have been achieved.
Quality of interaction: Assess the quality of discussions and exchanges of ideas.

2. Gathering feedback
Organise surveys or discussion groups to gather feedback from members on their experience within the network. Ask questions about :
Overall satisfaction.

Perception of the added value of the network.
Obstacles encountered and suggestions for improvement.

3. Analysis of concrete results
Evaluate the concrete results of the collaboration. This may include:
Completed projects: Number of projects launched and successfully completed.
Products or services developed: Innovations or improvements put in place thanks to the collaboration.
Established partnerships: New relationships or collaborations formed within the network.

4. Measuring Knowledge Sharing
Assess the mechanisms for sharing knowledge within the network. This can be measured by :
The number of shared resources (documents, case studies, etc.).
The frequency and quality of training courses or webinars organised.
Active use of knowledge-sharing tools.

5. Communications follow-up
Analyse the frequency and quality of communications within the network. This can include:
The number of messages exchanged on collaboration platforms.
The diversity of subjects covered.
The responsiveness of members to questions and discussions.

6. Assessment of Skills Developed
Assess whether members have acquired new skills through their participation in the network. This can be done by :
Members' self-assessment of their skills.
Organising tests or assessments after training courses.

7. Analysis of Participation Data
Use quantitative data to analyse participation. This can include:
The number of people attending meetings.
Member retention rate over time.
The frequency of engagement of active members.

8. Impact assessment
Measure the network's long-term impact on the members and communities it serves. This can include:
Changes in professional practices.
Performance improvements in members' projects or activities.
The economic or social impact of the initiatives launched by the network.

9. Periodicals
Organise periodic reviews to assess the progress made by the network. These reviews can include :
Quarterly or annual reviews.
Discussions on the successes and challenges encountered.
Adjusting strategies according to the results of the assessment.

03-What methods can improve knowledge sharing within the network?
Knowledge sharing is essential to maximise the effectiveness and impact of a collaborative network. Here are some key methods that can foster an environment conducive to knowledge sharing.

1. Collaboration Platforms
Using appropriate collaboration platforms (such as Slack, Microsoft Teams or Trello) makes it easier to share information. These tools allow members to discuss, share files and track projects in real time, making access to knowledge much smoother.

2. Webinars and Workshops
Organising regular webinars and workshops allows members to exchange ideas and learn from each other. These sessions can address specific topics, share case studies or feature guest experts, encouraging interaction and engagement.

3. Resource Library
Create a centralised resource library, accessible to all members, to store documents, case studies, practical guides and other useful materials. This library can be organised by theme to make it easier to find relevant information.

4. Mentoring and Pairing

Setting up mentoring or pairing programmes encourages the sharing of knowledge between experienced and new members. This encourages informal learning and the transmission of specific skills, while strengthening cohesion within the network.

5. Discussion Groups

Forming discussion groups or committees on specific topics allows members to focus on areas of common interest. These groups can meet regularly to exchange ideas, solve problems and share best practice.

6. Recognition systems

Establishing recognition systems to value contributions to knowledge sharing encourages members to participate actively. This can include awards, special mentions or leadership opportunities within the network.

7. Use of Social Media

Using social media to share knowledge, news and events allows you to reach a wider audience. Creating groups on platforms such as LinkedIn or Facebook can encourage interaction and engagement around specific topics.

8. Assessment and feedback

Encouraging members to give their opinion on knowledge-sharing methods and suggest improvements helps to adjust practices in line with the group's needs. Regular surveys can be set up to gather feedback.

9. Creating collaborative content

Encouraging the creation of collaborative content, such as blogs or newsletters, enables members to share their experience and knowledge. This creates a culture of sharing and gives everyone the opportunity to contribute.

10. Continuing training

Offering continuing education opportunities on relevant topics helps members to keep abreast of trends and innovations. This can include online courses, certifications or conferences, promoting skills development.

04-How can the effectiveness of knowledge-sharing methods be assessed?

Evaluating the effectiveness of knowledge-sharing methods is crucial to ensuring that they meet the needs of network members and contribute to the achievement of objectives. Here are a number of approaches and indicators to consider when carrying out this evaluation.

1. Define clear objectives
Before starting the evaluation, it is important to define clear objectives for knowledge sharing. These may include improving collaboration, increasing member engagement or disseminating specific information.
2. Gathering feedback
Organise surveys or focus groups to gather feedback from members on the methods used. Ask questions about :
Their satisfaction with sharing formats (webinars, documents, etc.).
The usefulness of shared information.
Obstacles to sharing knowledge.

3. Measuring Engagement
Analyse the level of engagement of members with sharing platforms and resources. This can include:
The number of participants in webinars and workshops.
Open and click rates on newsletters or e-mail communications.
The frequency of contributions on collaboration platforms.

4. Assessing the Quality of Interactions
Measure the quality of interactions within the network. This can be done by examining :
The depth of discussions in forums or newsgroups.
The diversity of subjects covered.
Members' responsiveness to questions and comments.

5. Analysis of concrete results
Evaluate the concrete results of knowledge sharing. For example, see if :
New projects have been launched thanks to the information shared.
Improvements have been observed in members' practices.

Collaborations and partnerships have been established as a result of these exchanges.

6. Monitoring Skills Developed

Assess whether members have acquired new skills as a result of sharing methods. This can be done by :

Member self-assessments before and after training courses or workshops.

Organising tests or assessments to measure learning.

7. Assessing Long-Term Impact

Measure the long-term impact of knowledge-sharing methods on members' practices. This may include:

Changes observed in project results or performance.

The sustainability of collaborations established through knowledge sharing.

The economic or social impact of the initiatives launched.

8. Periodic Reviews

Organise regular evaluations to review knowledge sharing methods. This can include:

Quarterly or annual reviews.

Discussions on the successes and challenges encountered.

Adjusting strategies according to the results of the assessment.

9. Comparison with Standards or Best Practice

Compare your knowledge sharing methods with industry standards or best practice. This can help identify gaps and areas for improvement.

Assessing the effectiveness of knowledge-sharing methods requires a systematic and multidimensional approach. By using a combination of qualitative and quantitative indicators, and by involving members in the process, it is possible to obtain a comprehensive overview of the performance of the methods and to identify opportunities for improvement.

-Conclusion: Towards Responsible Use of Pesticides

The responsible use of pesticides has become a crucial issue in the current context of global agriculture. Faced with the growing challenges of food safety, public health and preserving the environment, it is imperative to adopt practices that minimise the risks associated with these products. Raising awareness among farmers and consumers of the effects of pesticides is essential. A better understanding of the potential dangers can lead to more prudent and informed use.

Alternative methods, such as integrated pest management, offer sustainable solutions that reduce dependence on chemical pesticides. By integrating biological, cultural and mechanical strategies, farmers can effectively manage pests while preserving the health of their crops and the environment. In addition, the adoption of biological pesticides and less toxic formulations is a promising way of reducing the negative impact on human health and biodiversity.

Image : Notre Temps.com

Technological innovation also plays a key role in promoting the responsible use of pesticides. Precision tools such as drones and sensors enable farmers to apply pesticides in a targeted manner, minimising waste and reducing the risk of soil and water contamination. By providing real-time data on crop needs, these technologies help to optimise interventions and improve efficiency.

Public policies and regulations must support these changes by encouraging sustainable practices and limiting the use of the most harmful products. Setting strict standards for the registration and use of pesticides is essential to protect public health and the environment. This requires collaboration between governments, scientists and farmers to establish clear and accessible guidelines.

Farmer training is another key element in ensuring the responsible use of pesticides. Educational programmes need to be set up to inform farmers about alternatives, good application practices and the associated risks. The exchange of experience and knowledge within collaborative networks can reinforce this training and encourage the adoption of more sustainable methods.

It is also essential to involve consumers in this approach. Raising awareness of the effects of pesticides and the benefits of sustainably grown produce can influence purchasing choices and encourage farmers to adopt more responsible practices. Organic labels and short supply chain initiatives can also play an important role in this process.

Ongoing research is essential to develop new solutions and alternatives to chemical pesticides. Investment in research into biopesticides, resistant varieties and innovative farming practices can open up new prospects for sustainable agriculture. The results of this research must be disseminated and incorporated into farming practices to maximise their impact.

Finally, international collaboration is essential if we are to meet the global challenges posed by the use of pesticides. Exchanges of knowledge, best practices and innovations must be shared across borders to encourage a collective approach to these issues. Cooperation between countries, international organisations and players in the agricultural sector is crucial to building capacity and promoting responsible agriculture on a global scale.

In conclusion, responsible use of pesticides requires a combination of awareness-raising, training, technological innovation, regulation and collaboration. Every player, from farmers to consumers to decision-

makers, has a role to play in ensuring sustainable agriculture that respects human health and the environment. By adopting these approaches, it is possible to create an agricultural system that not only meets growing food needs, but also preserves our planet for future generations.

Sources and bibliographical references

Carson, R. (1962). Silent Spring. Houghton Mifflin.
Gurr, G. M., Wratten, S. D., & Luna, J. M. (2016). Biological Control: A Global Perspective. CABI.
Pardo, A., et al. (2018). "Impact of Genetically Modified Organisms on Biodiversity." Nature Sustainability, 1(1), 45-54.
Dill, G. M., et al. (2010). "Glyphosate-Resistant Crops: History, Status and Future." Pest
Management Science, 66(3), 307-318.
Fravel, D. R. (2005). "Commercialization of Biocontrol." Annual Review of Phytopathology,
43, 341-356.
Liu, Y. B., et al. (2005). "Bacillus thuringiensis: A Source of Insecticidal Proteins." Nature
Biotechnology, 23(3), 289-293.
Matsumura, F. (1985). Toxicology of Insecticides. Plenum Press.
Mason, G. J., et al (2004). "Rodenticides: A Review of Their Use and Impact." Wildlife Society Bulletin, 32(1), 92-96.
Alavanja, M. C. R., et al. (2003). "Pesticides and Cancer." Cancer Causes & Control, 14(2),
147-157.
Bogdan, A., et al. (2013). "Pesticide Residues in Food: A Review of the Health Risks." Journal
of Food Safety, 33(2), 203-215.
Kamel, F., et al. (2007). "Neurologic and Cognitive Function in Agricultural Workers."
Environmental Health Perspectives, 115(1), 103-108.
Matsumura, F. (1985). Toxicology of Insecticides. Plenum Press.
Raanan, R., et al. (2015). "Prenatal Exposure to Pesticides and Child Neurodevelopment."
Environmental Health Perspectives, 123(9), 991-997.
Baker, R., et al. (2013). "Long-term Ecological Effects of Pesticides." Environmental Science
& Technology, 47(12), 6357-6369.
Gilliom, R. J., et al. (2006). "Pesticides in the Nation's Streams and Ground Water, 1992-
2001." US Geological Survey Circular 1291.
Giller, K. E., et al. (2009). "Reducing Pesticide Use in Agriculture: A Review." Agriculture,

Ecosystems & Environment, 133(4), 243-251.
Gurr, G. M., Wratten, S. D., & Luna, J. M. (2016). Biological Control: A Global Perspective.
CABI.
Mineau, P., & Whiteside, M. (2013). "Pesticide Risk to Birds: A Review and Assessment of the
Evidence." Environmental Pollution, 173, 229-231.
Potts, S. G., et al. (2010). "Global Pollinator Declines: Trends, Impacts and Drivers." Trends in
Ecology & Evolution, 25(6), 345-353.
European Commission. (2018). "EU Action Plan for the Sustainable Use of Pesticides."
European Commission. (2019). "Regulation (EC) No 1107/2009."
EPA. (2021). "Pesticide Registration (PR) Notice." Environmental Protection Agency.
Baker, R., et al. (2013). "Long-term Ecological Effects of Pesticides." Environmental Science
& Technology, 47(12), 6357-6369.
European Commission. (2018). "EU Action Plan for the Sustainable Use of Pesticides."
Giller, K. E., et al. (2009). "Reducing Pesticide Use in Agriculture: A Review." Agriculture,
Ecosystems & Environment, 133(4), 243-251.
Gurr, G. M., Wratten, S. D., & Luna, J. M. (2016). Biological Control: A Global Perspective.
CABI.
Liu, Y. B., et al. (2005). "Bacillus thuringiensis: A Source of Insecticidal Proteins." Nature
Biotechnology, 23(3), 289-293.
Pretty, J. N., et al. (2006). "Resource-Poor Farmer Livelihoods in Developing Countries."
Agriculture, Ecosystems & Environment, 136(1), 1-9.
Alavanja, M. C. R., et al. (2003). "Pesticides and Cancer." Cancer Causes & Control, 14(2),
147-157.
Baker, R., et al. (2013). "Long-term Ecological Effects of Pesticides." Environmental Science
& Technology, 47(12), 6357-6369.
European Commission. (2018). "EU Action Plan for the Sustainable Use of Pesticides."
Giller, K. E., et al. (2009). "Reducing Pesticide Use in Agriculture: A Review." Agriculture,
Ecosystems & Environment, 133(4), 243-251.
Gurr, G. M., Wratten, S. D., & Luna, J. M. (2016). Biological Control: A Global Perspective.

CABI.
Kearsey, M. J., & Farquhar, A. (1998). "The Use of Plant Breeding to Improve the Resistance
of Crops to Disease." Journal of Agricultural Science, 130(4), 433-440.
Pretty, J. N., et al. (2006). "Resource-Poor Farmer Livelihoods in Developing Countries."
Agriculture, Ecosystems & Environment, 136(1), 1-9.
Baker, R., et al. (2013). "Long-term Ecological Effects of Pesticides." Environmental Science
& Technology, 47(12), 6357-6369.
Giller, K. E., et al. (2009). "Reducing Pesticide Use in Agriculture: A Review." Agriculture,
Ecosystems & Environment, 133(4), 243-251.
Gurr, G. M., Wratten, S. D., & Luna, J. M. (2016). Biological Control: A Global Perspective.
CABI.
Kearsey, M. J., & Farquhar, A. (1998). "The Use of Plant Breeding to Improve the Resistance
of Crops to Disease." Journal of Agricultural Science, 130(4), 433-440.
Liu, Y. B., et al. (2005). "Bacillus thuringiensis: A Source of Insecticidal Proteins." Nature
Biotechnology, 23(3), 289-293.
Pretty, J. N., et al. (2006). "Resource-Poor Farmer Livelihoods in Developing Countries."
Agriculture, Ecosystems & Environment, 136(1), 1-9.
Alavanja, M. C. R., et al. (2003). "Pesticides and Cancer." Cancer Causes & Control, 14(2),
147-157.
Baker, R., et al. (2013). "Long-term Ecological Effects of Pesticides." Environmental Science
& Technology, 47(12), 6357-6369.
European Commission. (2018). "EU Action Plan for the Sustainable Use of Pesticides."
Giller, K. E., et al. (2009). "Reducing Pesticide Use in Agriculture: A Review." Agriculture,
Ecosystems & Environment, 133(4), 243-251.
Gurr, G. M., Wratten, S. D., & Luna, J. M. (2016). Biological Control: A Global Perspective.
CABI.
32
Kearsey, M. J., & Farquhar, A. (1998). "The Use of Plant Breeding to Improve the Resistance
of Crops to Disease." Journal of Agricultural Science, 130(4), 433-440.
Pretty, J. N., et al. (2006). "Resource-Poor Farmer Livelihoods in Developing

Countries."
Agriculture, Ecosystems & Environment, 136(1), 1-9.
Alavanja, M. C. R., et al. (2003). "Pesticides and Cancer." Cancer Causes & Control, 14(2),
147-157.
Baker, R., et al. (2013). "Long-term Ecological Effects of Pesticides." Environmental Science
& Technology, 47(12), 6357-6369.
European Commission. (2018). "EU Action Plan for the Sustainable Use of Pesticides."
Giller, K. E., et al. (2009). "Reducing Pesticide Use in Agriculture: A Review." Agriculture,
Ecosystems & Environment, 133(4), 243-251.
European Food Safety Authority (EFSA). (2021). "Annual Report on Pesticide Residues in
Food."
Giller, K. E., et al. (2009). "Reducing Pesticide Use in Agriculture: A Review." Agriculture,
Ecosystems & Environment, 133(4), 243-251.
Alavanja, M. C. R., et al. 2003. "Pesticides and Cancer." Cancer Causes & Control, 14(2),
147-157.
European Commission. "Farm to Fork Strategy.
USDA. "Integrated Pest Management (IPM) Principles.
Ministry of Agriculture and Food Sovereignty, France. "Bees and Pesticides Project."
California Department of Food and Agriculture. "Healthy Soils Initiative.
FiBL & IFOAM. "The World of Organic Agriculture.
BUND (2021). "Pesticides - What You Need to Know."
EPA (2021). "Pesticide Residues and Your Health.
European Commission (2020). "Eco-Schools: A Sustainable Schools Initiative.
Ministerio de Agricultura (2020). "Awareness campaigns on organic products.
Ministry of Agriculture (2022). "Training courses on the responsible use of pesticides.
UC Davis (2021). "Research on Pesticides and Public Health.
European Commission. (2021). "EU Action Plan for the Sustainable Use of Pesticides."
Garrido, F. et al. (2018). "Improving the Safety of Systemic Pesticides." Journal of Pesticide
Science.
Gould, F. (2020). "Sustainable Management of Resistance to Pesticides." Annual Review of
Entomology.
Hajek, A. E. et al. (2016). "Biological Control: A Global Perspective." CABI.
Isman, M. B. (2020). "Pesticide Resistance and the Future of Bio-Insecticides."

Pesticide
Biochemistry and Physiology.
James, C. (2019). "Global Status of Commercialized Biotech/GM Crops." ISAAA.
Khan, Y. et al. (2020). "Nanopesticides: A New Approach to Sustainable Agriculture."
Agricultural Sciences.
Weber, D. C. et al. (2020). "Integrated Pest Management: A Sustainable Approach.
Pest
Management Science.
Zhang, Y. et al. (2021). "Precision Agriculture Technologies for Pest Management."
Agronomy
Journal.

Contents

Farmers are the first to be affected by these risks. What exactly are they?

The collective report confirms that there is a strong presumption of a link between occupational exposure to pesticides and seven diseases: multiple myeloma (a blood cancer), prostate cancer, Parkinson's disease, cognitive disorders, non-Hodgkin's lymphoma (another blood cancer), chronic obstructive pulmonary disease and chronic bronchitis. Links are suspected for other diseases, but the evidence is less strong. These include Alzheimer's disease, anxiety disorders, certain cancers affecting the bladder or kidney, asthma and wheezing, and thyroid disease.

What are the risks when exposure to pesticides occurs in utero or during early childhood?

Here too, there are numerous studies, particularly epidemiological studies, which suggest a strong presumption of a link between exposure in utero or during childhood and the risk of certain cancers, particularly leukaemia and central nervous system tumours, which are the most common paediatric cancers. In addition, a number of studies suggest that exposure to certain insecticides increases the risk of behavioural disorders and impaired motor, cognitive and sensory functions in children.

What about people living near agricultural areas?

It is much more difficult to determine a level of risk with strong evidence. In fact, these people may be contaminated by substances spilled on crops, but this level of contamination is much more difficult to assess precisely. However, there is a suspected link between the exposure of people living near farmland and Parkinson's disease. Similarly, certain behaviours suggestive of autism spectrum disorders in children could be linked to residential proximity to pesticide spraying zones (less than 1.5 km).

Pesticides: what are we talking about?

Pesticides are plant protection products used in agriculture to combat either plant species considered undesirable in crops, or harmful organisms. These products, whether they are still authorised today or have been banned but remain in the environment, give rise to numerous concerns about their possible effects on human health and, more broadly, on biodiversity.

Source : https://www.frm.org/fr/actualites?

Biodiversity and the services provided by nature: what do we know about the impact of pesticides?

To what extent are birds, insects and other living organisms affected by pesticides? What effect do these substances have on services as essential as pollination or biological pest control? A joint INRAE-Ifremer scientific report provides up-to-date information on the impact of plant protection products on biodiversity and on the services that ecosystems provide to society. It identifies ways of reducing them. The result of two years' work by a multi-disciplinary team, it was commissioned by the Ministries of the Environment, Agriculture and Research, and covers all environments: air, land, freshwater and marine.

Published on 16 May 2022

Why this collective scientific expertise?

The decline in biodiversity in France and around the world has been documented in an increasing number of publications over the last twenty years, a warning reinforced by the publication of the 2019 IPBES report[1]. Environmental contamination by pesticides has been identified as one of the causes of this situation. Ambitious public policies are now in place in France and Europe to regulate the use of plant protection products - in other words pesticides used on crops, gardens, green spaces and infrastructure - and to encourage alternative solutions to synthetic pesticides. But the use of these substances remains significant: 55,000 to 70,000 tonnes are still used in France, depending on the year. Public authorities, farmers, environmental protection associations, water agencies, scientists, manufacturers, citizens... all are concerned by this issue, which affects our environment and our food. To obtain an up-to-date overview of scientific knowledge, the Ministries of the Environment, Agriculture and Research asked INRAE and Ifremer to conduct a collective scientific assessment of the impact of plant protection products on biodiversity and ecosystem services. The results of this multidisciplinary and collective exercise, which involved 46 experts from 19 research bodies over a 2-year period, were shared at a feedback symposium attended by over 600 participants on 5 May.

In 2005, the public authorities commissioned INRA and Cemagref (now merged into INRAE) to carry out an initial assessment collective scientific of pesticides and their impact on agriculture and the environment, with the effects on human health being assessed by INSERM. In 2008, INRA delivered another report on the impact of agriculture on biodiversity. These expert reports have helped to make pesticide reduction a key research priority, and were instrumental in the development of France's first plan to reduce the use of pesticides in agriculture. Launched in 2008 following the Grenelle Environment Forum, the Ecophyto 2018 plan set an ambitious, quantified target: to halve the use of pesticides in 10 years. Since then, the regulations governing the marketing of pesticides have evolved, with the withdrawal of substances with the highest persistence in the environment or the highest toxic effects, the recognition and prioritisation of biocontrol, and the introduction of a phytopharmacovigilance[2] system to monitor the effects of pesticides during their use. In addition, the European Union has adopted the Water Framework Directive and the Marine Strategy Framework Directive, or the pesticides package, to better protect human health and the environment against the risks associated with chemical substances. Agricultural practices are changing, with greater emphasis on biocontrol solutions, organic farming and agro-ecology. Scientific knowledge and analytical techniques have improved, enabling us to better characterise the effects of substances and the contamination of ecosystems. They are also mobilising new concepts: ecosystem services, i.e. services that nature provides to society (such as pollination or biological pest control), holobionts formed by an organism, e.g. a plant, and the communities of micro-organisms that are closely associated with it, known as its microbiota), exposome (all the chemical, physical and biological exposures to which we are subjected), etc.

Based on a literature review of more than 4,000 international scientific publications, the report published on 5 May by INRAE and Ifremer provides an update on the current state of knowledge, highlighting what is solid, still to be consolidated or insufficiently explored since the previous reports. This data is essential for French and European public authorities, in terms of regulations and research strategy. In addition to this INRAE-Ifremer report, Inserm has updated its report be published in 2021on the impact of pesticides on human health, to , and INRAE has been commissioned to produce a report on the management of plant cover to promote the natural regulation of crop pests (results to be delivered on 20 October).

What lessons can be learned about impact?

On land and at sea:

plant protection products are present everywhere

The picture drawn in 2022 is much more accurate than those drawn in 2005 and 2008, thanks to a denser monitoring network and improved analytical techniques for plant protection products and certain products derived from their degradation (or transformation). Glyphosate - by far the most widely used herbicide in the world - and its main transformation product (AMPA) are among the most widespread substances found in soils. All environments are affected by the contamination of mixtures of plant protection products, but agricultural areas close to application sites are the most contaminated. This contamination then affects soils and watercourses as far as the seas and oceans, with concentrations generally decreasing along this continuum. Some pollutants that are very persistent in the environment, such as DDT, lindane and hexa-chlorobenzene, which have been banned for years, can even be found in the deep seas and polar zones. Mixtures of pollutants are found everywhere: phyto-pharmaceutical products, themselves in mixtures, are present alongside medicines, microplastics....

On the other hand, little scientific data is available on contamination in the French overseas territories, which have their own specific characteristics linked to particular regulations. One exception is chlordecone, which was authorised under a derogation until 1993 (after being banned in 1990) in the banana plantations of the French West Indies. It has been the subject of numerous studies in Guadeloupe and Martinique, where it is present along the entire land-sea continuum, although it becomes more diluted as one moves away from the spraying zones.

Biodiversity and the services it provides are affected

Chemical pollution - to which plant protection products contribute - appears to be the third or fourth factor in the destruction of biodiversity worldwide, behind the change or destruction of natural habitats, the exploitation of resources and climate change. Current knowledge of the specific effects of plant protection products is mainly generated in agricultural contexts, the main products studied being synthetic pesticides and copper. Very little is known about biocontrol products. These are specific to French regulations, and include macro-organisms (insects, nematodes, etc.) and micro-organisms (viruses, bacteria, fungi, yeasts, etc.) introduced into crops to combat their pests, chemical mediators (such as pheromones or kairomones) that trap, divert and disrupt attackers, and natural substances. The latter are of mineral, plant or animal origin and have a variety of uses (fungicides, insecticides, herbicides, etc.).

There is strong evidence that, in agricultural areas, plant protection products are one of the main causes of the decline in terrestrial invertebrates, including insect pollinators and pest predators (ladybirds, carabid beetles, etc.), as well as birds. In granivorous birds, direct effects are predominant, due to the toxicity of the seeds ingested. For insectivorous birds, the indirect effects are major: they lose their food supply because of the reduction in the number of insects. Aquatic organisms are also affected.

Macroinvertebrate populations could fall by 40% in the most polluted agricultural waterways. For all these terrestrial and aquatic organisms, the direct and indirect non-lethal effects are major, something that was not studied to the same extent 15 years ago. These effects can take the form of loss of orientation or flight capacity in insects and birds, reduced reproductive efficiency or immune deficiencies.

Some are due to damage to the microbiota, i.e. all the micro-organisms present in these organisms (like the intestinal flora in humans, for example). Some of these products also have effects on the endocrine system. These effects are mainly being studied in model species of mammals and fish, and their impact on populations has yet to be assessed.

Plant protection products are also responsible for the decline of amphibians (25% of their populations are threatened in Europe) and bats. In addition to these findings by type of organism, the study also looked at the ecological role they play.

Microorganisms, which are present in abundance in all environments, are mainly affected in agricultural soils and nearby waterways contaminated by plant protection products. Their capacity to break down organic matter and supply nutrients to ecosystems is diminished.

The report highlights the importance of landscape management for the resilience of biodiversity: it is essential to create refuge areas (terrestrial and aquatic vegetation) that are connected to each other: they are reservoirs of species that can then recolonise neighbouring environments. However, these refuge areas, and more generally natural habitats, are threatened by the simplification of agricultural landscapes (larger plots of land with less varied borders) and by soil artificialisation, with plant protection products playing an aggravating role.

The three ecosystem services for which the impact of pesticides has been studied the most to date are crop production (protected from pests and diseases by plant protection products), pollination (negatively affected, mainly by neonicotinoid and pyrethroid insecticides) and the control provided by natural predators against crop pests (also negatively affected).

However, the latter two services are useful to the former: plant production will therefore also be negatively impacted in the long term. There are many gaps in our knowledge of other services.

What levers can be used to limit these impacts?

In addition to reducing the use of pesticides, which remains the primary lever for preserving biodiversity, the report identifies three main types of action: mitigation of effects, regulation, and the use of less persistent and less impactful products such as biocontrol products.

Mitigating the effects

Mitigating the effects can involve limiting the drift (off target) of phytopharmaceutical products during application and their immediate transfer to the soil and water, which means taking action when the weather is favourable (now regulated). Appropriate farm equipment and products with heavier formulations can help to limit drift. It is important to cover the soil with vegetation: 40% less polluted in this case. Phyto-remediation - or decontamination by plants - can help to reduce the presence of pesticides in soil by 10%. Landscape management also plays a major role, creating buffer zones that limit run-off into neighbouring soils and watercourses: ponds can reduce run-off by 60%, while hedges and grassed areas can reduce it by 40%.

A complementary approach is to favour biocontrol over synthetic pesticides. However, there is a need for research and data on environmental contamination and its impact. In addition, the difficulty is that some biocontrol products and organisms are already present in the environment, and it is difficult to distinguish between the fraction contributed by phytopharmaceutical control. Biocontrol solutions are generally less persistent and have lower ecotoxicity than conventionally used pesticides, although there are exceptions such as Bacillus thuringiensis, spinosad and pyrethrins.

There is no single solution for reducing the impact of pesticides, but a need to combine different levers for greater effectiveness.

The regulations

Current regulations, although very ambitious and restrictive, as highlighted by the scientific literature, could be improved to better achieve their protection objectives. At present, routine risk assessment does not take into account the complexity of living organisms (and interactions between species) and cocktail or multi-stress effects. Improvements could also be achieved by integrating epidemiological monitoring data to a greater extent. We also need to better target the species studied (e.g. granivorous birds to better assess indirect effects) and the environments studied.

Concluding his report, François Houllier, CEO of Ifremer, affirmed his institute's commitment to pursuing research to shed more light on the gaps: "Most of the levers are on land, but we need to know more about the sea: The Blue Chair set up by Ifremer in Nantes will study the impact of pollution on the microbiota of marine organisms; I also hope that ambitious projects will be submitted, on the marine exposome in the "Ocean - Climate" PPR, on the land-sea continuum in the "One Water" PEPR, or as part of the European mission for the restoration of oceans and waters". For his part, Philippe Mauguin, Chairman and CEO of INRAE, stressed the importance of strengthening expertise and developing alternative approaches. The "One Health" concept - a single approach to human, animal and environmental health - could be further developed through research into the concepts of exposome, ecoexposome and microbiota. We need to make progress in the qualitative and quantitative characterisation of biodiversity and ecosystem services, and in the attribution of the multiple causes of decline: chemical pollution, climate change, land artificialisation, etc. We also need to correct the magnifying glass effect created by focusing research on particular molecules or chemical groups, and the reverberation effect whereby a result can only be obtained if it is present in the area 'lit up' by the research.

The public authorities have mobilised significant resources to support alternatives to pesticides, in particular those managed by INRAE: the priority research programme Cultiver et protéger autrement (Growing and protecting differently), the priority research programme and equipment (PEPR) on plant breeding, and the major challenges of biocontrol and robotics. Agroecology and alternatives to pesticides are a priority in INRAE's 2030 roadmap, and INRAE has also been the driving force behind a European alliance to phase out pesticides, bringing together 34 research institutes from 20 European countries. A symposium organised by INRAE in Dijon on 2 and 3 June, under the aegis of the French Presidency of the European Union, will provide an opportunity to share advances and research avenues in this field. In addition to the French public authorities, Philippe Mauguin stresses the need to raise awareness among the European public authorities and to share this expertise in an operational way. This was done on 11-12 May at a ministerial conference organised by the French Ministry for the Environment, when Anses presented the European Partnership for Risk Assessment of Chemical Substances (PARC), which it is coordinating, with the aim of designing a new generation of more protective risk assessment. In December 2022, INRAE will also be presenting an outlook on the phasing out of pesticides at European level.

Notes :

1. *IPBES: Inter-governmental Science-Policy Platform on Biodiversity and Ecosystem Services.*
2. *Phytopharmacovigilance: System designed to monitor the undesirable effects of plant protection products available on the market, covering environmental contamination, exposure and impacts on living organisms - including humans - and ecosystems, as well as the emergence of resistance (Anses).*

Source : https://www.inrae.fr/

Preventing risks linked to pesticides
By: G.L/ABU YOCEF/2023

Preventing risks linked to pesticides
2023.

MIX
Papier aus verantwortungsvollen Quellen
Paper from responsible sources
FSC® C105338

Printed by Books on Demand GmbH, Norderstedt / Germany